God Created

DOGS

"In the beginning God created..." Genesis 1:1

God Created Dogs

First printing: June 2024

Designed by Melissa Marquez
Written by the ICR Communications team: Jayme Durant, Beth Mull, Michael Stamp, Lori Fausak, Bethany Trimble, Renée Dusseau, and Rachel Brown

ISBN (paper): 978-1-957850-66-5
ISBN (ebook): 978-1-957850-69-6

All Scripture quotations are from the New King James Version.

Please visit our website for other books and resources: ICR.org

Printed in the United States of America.

Cover Image: Siberian Husky

Published by Creation Kids Books, an imprint of ICR Publishing Group

ICR PUBLISHING GROUP

God Created

DOGS

ICR
INSTITUTE FOR CREATION RESEARCH

Creation Kids Books

Dallas, TX
ICR.org

LET'S RAISE THE WOOF!

There's so much to dig up about dogs! For centuries, these furry friends have been loyal, lively, and loving companions to humans. Dogs live all over the world, and they're cherished by families in many different cultures. It's no wonder they're known as "man's best friend"!

Most importantly, dogs show God's special design for His creation. Their natural skills and instincts make them great helpers, protectors, and playmates. What can we learn about these captivating canines?

DID YOU KNOW?

Dogs are the most common pet in the United States. It's estimated that dogs are found in over 65 million households around the nation. Talk about paw-pular!

CANINE CREATION

God created dogs on Day 6 of the creation week. We don't know what the first dogs looked like, but they were part of the canine family—a group that includes foxes, coyotes, and others. The Lord Jesus gave dogs a huge genetic potential, so there are now hundreds of breeds!

Red Fox

Arctic Wolf

SIX DAYS OF CREATION

DAY 1

DAY 2

DAY 3

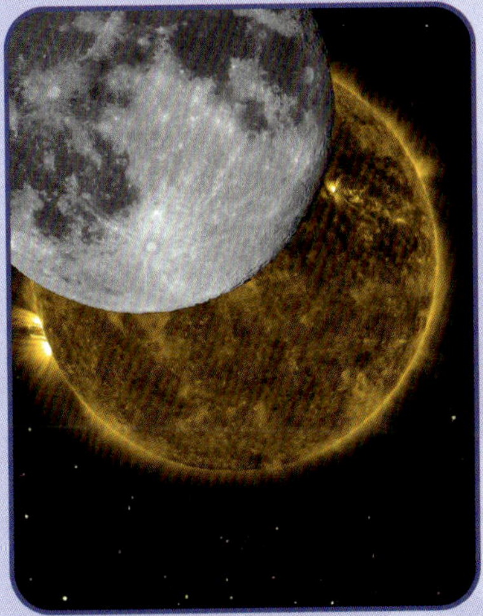

DAY 4

DAY 5

DAY 6

RE-BARK-ABLE SENSES

Dogs are some of the world's nose-iest animals—they know a lot about someone based on scent. With their expert sense of smell, dogs can identify diet, feelings, and even certain health problems. In fact, they smell over 60 times better than humans do. Dogs use 40 times more of their brain to analyze odors, too!

Not only that, but dogs also have incredible hearing. Their ears can pick up a wide range of low- and high-pitched sounds that human ears can't. And they can hear noises four times farther away than we can.

Ever wondered why a dog's nose is always wet? Dogs naturally produce mucus to help them collect important scent particles. They also lick their noses to keep themselves cool.

DID YOU KNOW?

While dogs can't see all the colors that humans can, they see colors of gray, yellow, and blue. Their night vision is spectacular, too!

German Shepherd

Bloodhound

WHO LET THE DOGS OUT?

Dingo

Bengal Fox

The better question might be, who let them in? All kinds of canines have lived in the wild for thousands of years. The dogs we adopt into our families are special because they're domesticated, or bred to live with humans.

But several of their untamed relatives are still found across the world today. God designed canines so they could thrive in a variety of places. Check out these fur-ocious animals!

Gray Wolf

Domestic dogs and wolves are different in many ways. Wolves have larger paws, stronger jaws, and straighter tails than their tame cousins. Their eyes are always yellow- or amber-colored, while other dogs' eyes come in a range of hues. They're usually shy and avoid humans, too.

African Wild Dog

Dhole

HERDING DOGS

The American Kennel Club puts dogs into seven primary groups: herding, working, sporting, non-sporting, hound, terrier, and toy. Herding dogs are famous for their ability to herd animals like sheep, goats, and cattle. These energetic canines are super intelligent and fast thinkers. They help farmers by guiding and protecting livestock. They're incredibly loyal dogs that enjoy staying active.

Old English Sheepdog

Cardigan Welsh Corgi

Border Collie

"The LORD is my shepherd; I shall not want. He makes me to lie down in green pastures; He leads me beside the still waters. He restores my soul; He leads me in the paths of righteousness for His name's sake." (Psalm 23:1–3)

Australian Shepherd

WORKING DOGS

Working dogs are often trained for heavy-duty tasks like protection and rescue missions. Some pull sleds or carts. Others find lost people or work with police and the military as detectors, trackers, or scouts. Their created instincts and skills are particularly helpful for the positions they hold. While these jobs are important, working dogs can also be great protectors and pets for their human families.

Rottweiler

Great Dane

DID YOU KNOW?

Dogs have two kinds of sweat glands, but only the ones in their paws cool them down. That's why they pant when they're hot.

Saint Bernard

SPORTING AND NON-SPORTING DOGS

Sporting dogs were originally used to help hunters find and retrieve wildfowl, or birds. Many of them have water-resistant fur coats that can withstand extreme conditions. These canines often love to swim and exercise with their humans.

When it comes to non-sporting dogs, the name isn't actually accurate. Many of these pooches like to be active, too! This group is a mix of dog breeds that don't really fit into any of the other six categories. Make no bones about it—they're all lovable pups that have made people's lives better.

Boston Terrier

Dalmatian

HOUND DOGS

Like sporting dogs, hounds were bred to help humans in their hunting expeditions. They're renowned for their top-notch eyesight and ability to detect and follow scents. While hounds are dedicated to their work, these independent, quiet canines are often known to be couch potatoes.

Afghan Hound

Beagle

Basset Hound

Dachshund Puppies

Is a dog's licking the same as kissing? Maybe. It could mean that your dog is happy to see you—or, if you just finished eating, to let you know that he/she wants a taste! Human skin also contains natural salt, so that might be why your dog likes to lick you, too.

TERRIERS AND TOYS

What these small canines lack in size, they make up in personality! Terriers are enthusiastic dogs. They were bred to keep rats and other rodents away. Short-legged terriers are small enough to follow pests into tight places, while long-legged terriers have the features to dig them out.

Toy dogs may be tiny, but they have big, loving hearts. They're perfect dogs for urban life since they're little enough for apartments and small yards. But their favorite place to be? Sitting on your lap!

West Highland White Terrier

Pug Puppies

But what about mutts? Mutts are always one-of-a-kind because they're a unique mix of dog breeds! Since mutts generally have fewer genetic health issues, they tend to live longer.

Chihuahua

ROYAL BREEDS

Throughout history, dogs have served as beloved companions to people from many different backgrounds—including royalty! These treasured pooches have lived alongside nobility since ancient times.

Queen Elizabeth II popularized Pembroke Welsh corgis. She had more than 30 corgis during her reign. What dogs have captured the hearts of other kings and queens?

The royalty of greyhounds can be traced back to the Egyptians. Only nobles were allowed to own them. When they died, these dogs were mummified and buried alongside their owners.

Cavalier King Charles spaniels were named after King Charles II. They served as part of European court life in the seventeenth century.

King George VI was known for owning yellow Labrador retrievers, which weren't an officially recognized breed in England until 1903.

UN-FUR-GETTABLE FRIENDS

The Lord Jesus created dogs with all kinds of stylish coats. From fluffy to furless, each type serves a unique purpose. Some dogs, like huskies, have thick coats with two layers. Huskies were bred to pull sleds through the snow, so their thick fur helps keep them warm and protects their skin. In 1925, a group of Siberian huskies became famous for sledding 674 miles across Alaska in -40°F weather to deliver a life-saving serum.

Husky

Komondor

DID YOU KNOW?

Dogs shed old or damaged fur so it can be replaced with new, healthy hair.

Samoyed

Australian Cattle Dog

DOGS AROUND THE WORLD

The American foxhound is one of the oldest breeds in the United States.

Chihuahuas are from Mexico. These tiny dogs have big personalities.

DID YOU KNOW?

The pharaoh hounds of Malta got their name because it was believed they descended from the dogs shown on wall paintings in ancient Egyptian tombs.

Golden retrievers are from England, and they love to swim.

Akitas originated in northern Japan, where they were used for guarding and hunting.

BOW TO THE WOW!

Isn't it amazing how God designed all the extraordinary features dogs have today? These loyal animals have blessed humans in countless ways. From wet noses to wagging tails and everything in between, there are so many things to love about these wonderful canines. Dogs are paw-sitively delightful friends that reflect the goodness of their Creator, the Lord Jesus Christ.

Ruff, woof, arf, yip, yap, bow wow! The words used to describe how a dog barks are based on the breed and the language of the person who hears it. What do you think dogs sound like?

Spanish: guau guau

Greek: ghav ghav

German: wau wau, wuff wuff

French: ouaf ouaf

Japanese: wan wan, kyan kyan

Mandarin Chinese: wang wang

Italian: bau bau

Russian: gav gav

Arabic: hau hau

"The LORD is good to all, and His tender mercies are over all His works." (Psalm 145:9)

ICR DOGS

Macey

Chocolate Labrador Retriever

Luna

German Shepherd

Pablo

Chihuahua Jack Russell Terrier Mix

Sunny

German Shepherd Mix

Calvin

Beagle Shih Tzu Mix

Augustine

Goldendoodle

At the Institute for Creation Research, we love our furry BFFs! These steadfast companions are family—and the inspiration for this book. Here are a few pictures of the dogs that leave pawprints on the hearts of our staff.

Cocoa

Staffordshire Bull Terrier

Niki

Golden Retriever

Frida

Shih Tzu Maltese Mix

Tucker

Shih Tzu Poodle Mix

Colonel

Great Pyrenees

Pippin

Pembroke Welsh Corgi

There are lots of different breeds pictured here...and countless more to discover! What types of dogs can you find?

FIND OTHER CHILDREN'S RESOURCES AT ICR.org/store

Image credits (t: top; b: bottom; l: left; m: middle; r: right)

Bigstock Photos: cover, 1, 3-6, 8-9, 10b, 11t, bl, 12, 14-20, 21t, 22b, 23b, 24-26, 27t, bl, 28-29

iStockPhoto: 7tl, tm, 10t, 11br, 13, 21b, 22t, 23t, 27br

Adobe Stock/Fotolia: 7tr, bm, br

NASA: 7bl

ICR: 30-31